TAGGING BATS

By Kaitlyn Desrochers, Kennedi Emery, and Brynn Burnsworth
with Dr. Yossi Yovel

Dedicated to the researchers and developers of technology who are working together to answer the most fascinating questions of our time.

Text by Kaitlyn Desrochers, Kennedi Emery, Brynn Burnsworth, with contributions from Dr. Yossi Yovel and Dr. Ellen Cavanaugh

Images for the cover and pages 4, 6, 7, 8, 9, 11, 16, 17, 18, 22, 23, 24, and 29, from yossiyovel.com with credit to Eran Amichai, Eran Levin, Jens Rydell, Mickey Samuni-Blank, Ofri Eitan, Ivo Borissov and many other lab members of the Bat Lab for Neuro-Ecology. Images of the Vesper Senser provided by Joanna Weinstein. Used with permission.

Cover design by Marilyn Freeman

Copyright © 2018 by Grow a Generation

All rights reserved. This book or any portion thereof may not be reproduced or used in any manner whatsoever without the expressed written permission of the publisher except for the use of brief quotations in a book review or scholarly journal.

ISBN 978-0-359-44248-5

Grow a Generation
Sewickley, PA 15143
www.growageneration.com

Any and all profits from the sale of this book benefit Bat Conservation International

This is a Dragon Tag. Students at our school in Western Pennsylvania wear them on their backpacks. The Dragon Tags help students and adults to know what bus a student should be on, or if a parent is picking the child up from school.

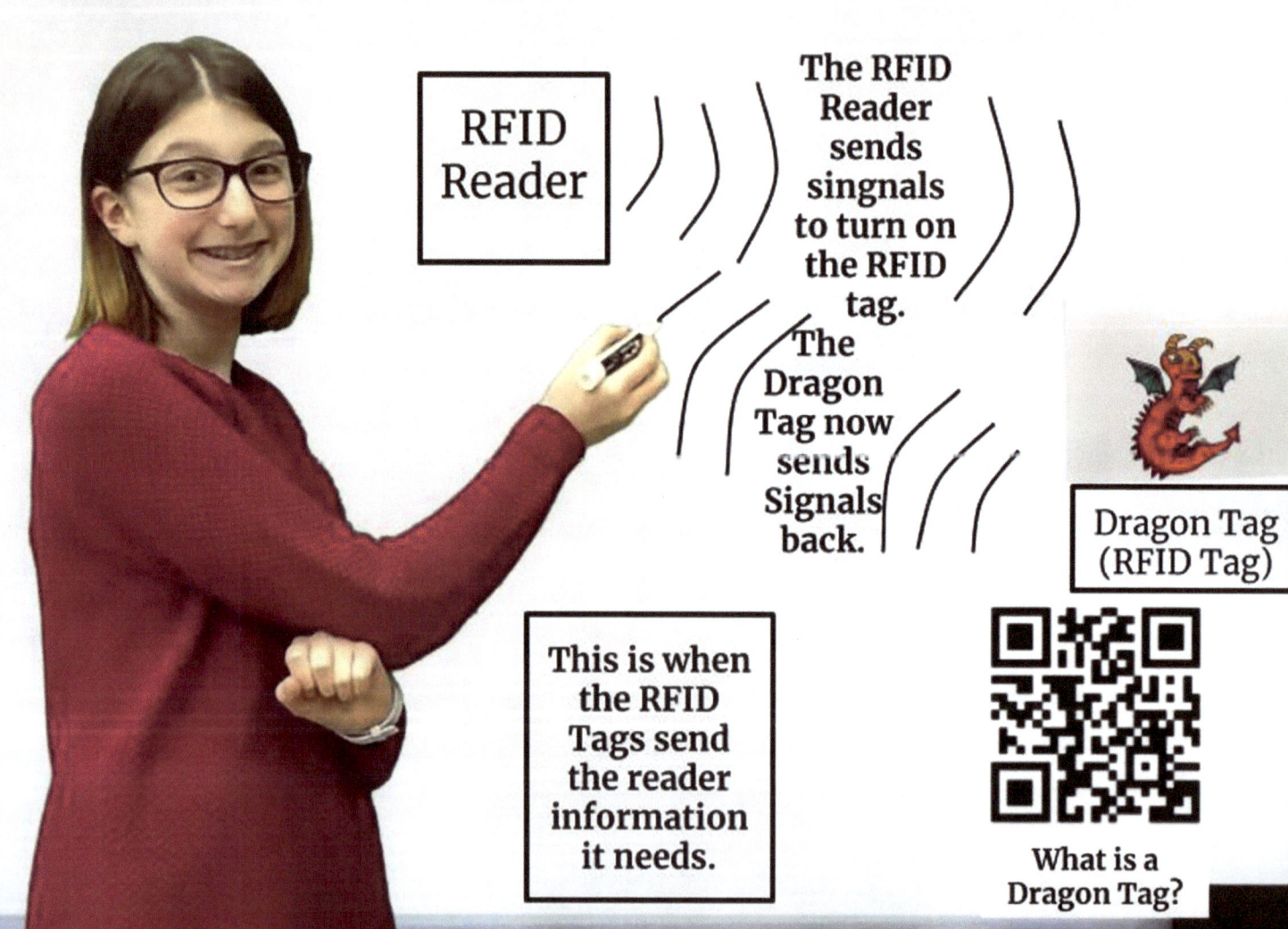

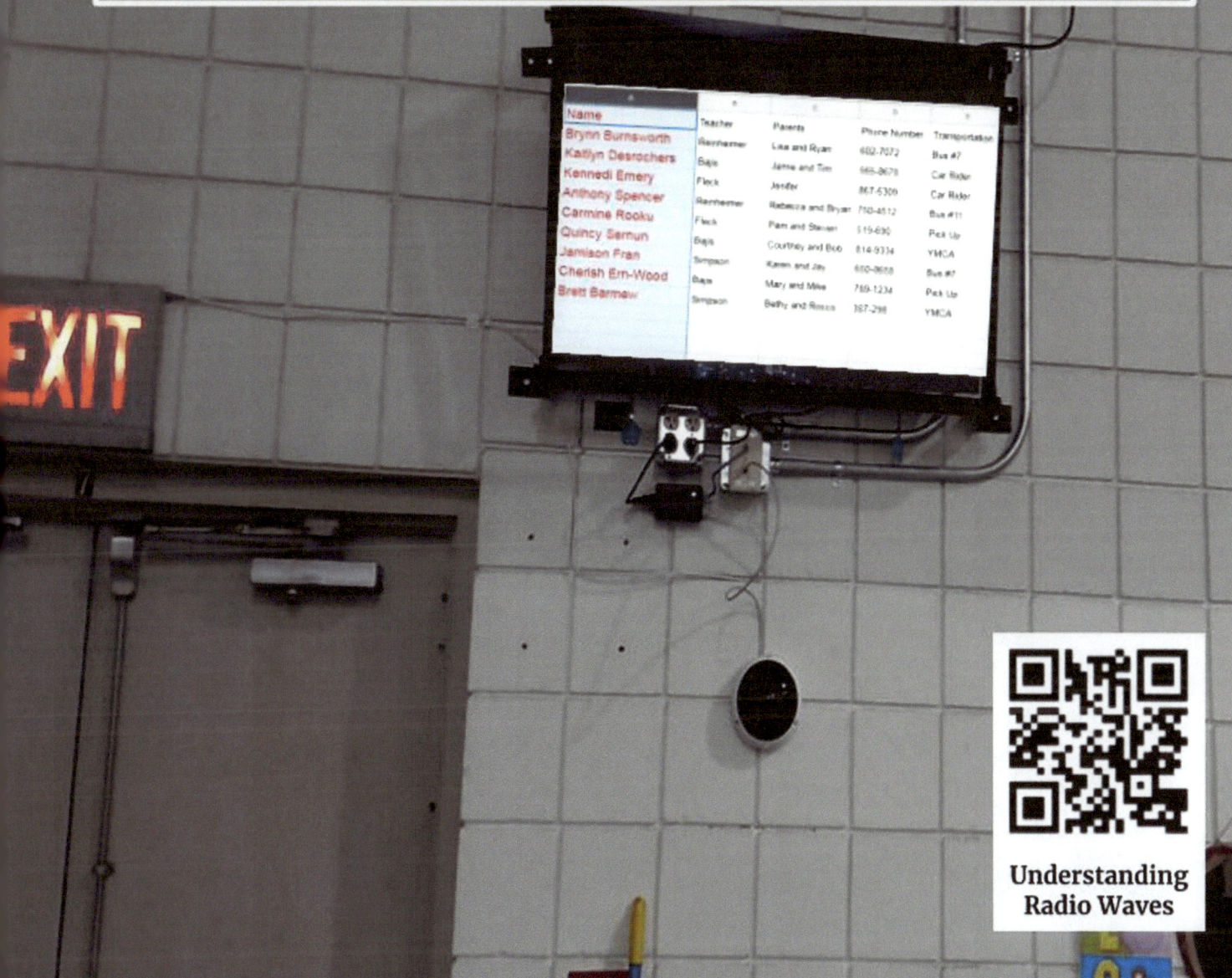

The code transmits through the RFID tag reader to a computer. Then, the computer takes the code and displays all of the student information on the screen. It is also saved to a computer file for access at a later time, if needed.

Understanding Radio Waves

We are not the only ones who use RFID! Recently, we learned of the amazing Bat Lab of Dr. Yossi Yovel, all the way around the world in Israel! His lab uses all types of new technology to study bats and how they think and communicate.

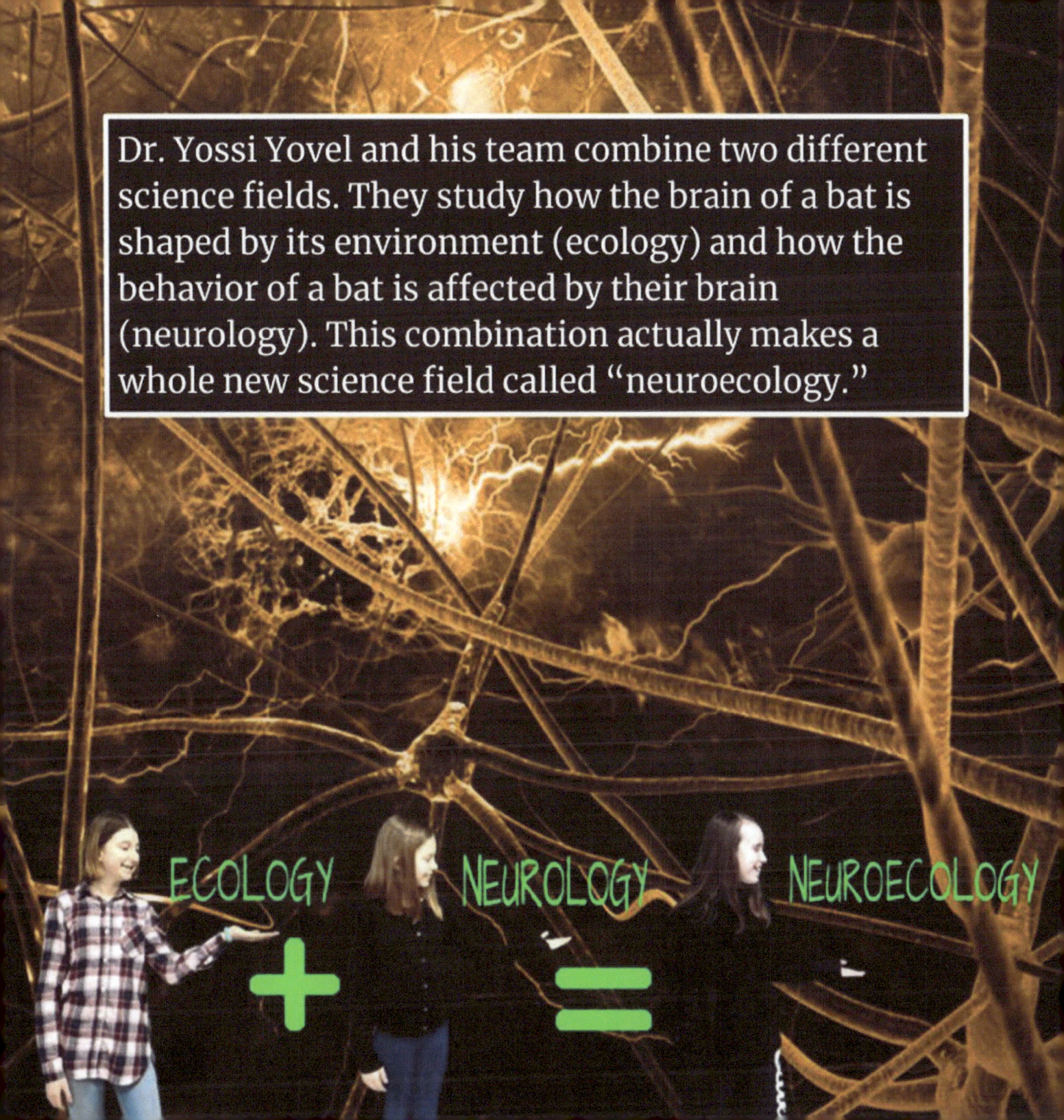

Dr. Yossi Yovel and his team combine two different science fields. They study how the brain of a bat is shaped by its environment (ecology) and how the behavior of a bat is affected by their brain (neurology). This combination actually makes a whole new science field called "neuroecology."

ECOLOGY + NEUROLOGY = NEUROECOLOGY

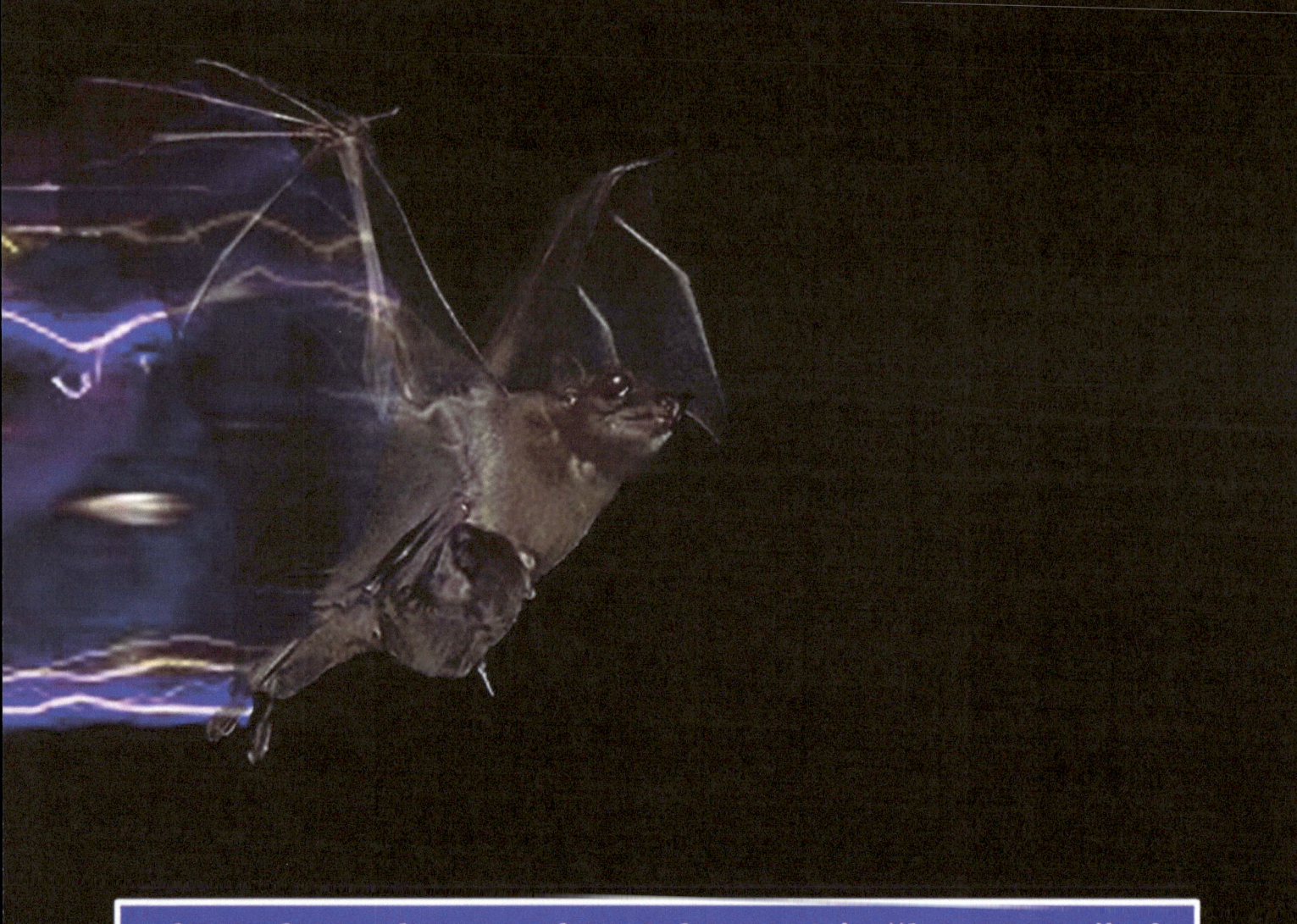

Where does a bat go when it leaves it's "hang out." If you have ever tried to watch a bat fly, you know it's hard to follow them with your eyes. They can fly at speeds of up to 40 miles per hour and can quickly change direction mid-flight. Dr. Yovel and his team needed technology to help track bats.

Dr. Yovel's lab worked with engineers at Vesper and made something they call the Vesper Sensor. They study Egyptian Fruit Bats. Each bat wears the Vesper Sensor like a bookbag. The sensor stays on the bat as they fly.

Vesper Sensors are so cool! They are actually a combination of 4 different sensors, all amazing in themselves!
- GPS
- Vesper Microphone
- 9DOF Sensor (Compass, Gyroscope, and an Accelerometer Combo Pack)
- Electrophysiology Sensor

This is a Vesper Sensor Compared to the size of a U.S. Quarter.

Go ahead, call us nerdy, but we find it fascinating to actually understand what a computer sensor can see. Think about it...we can now "see" and measure things our human eyes and senses could never perceive. We can also record echolocation to monitor what the bats are "seeing," and monitor their social calls, or bat language.

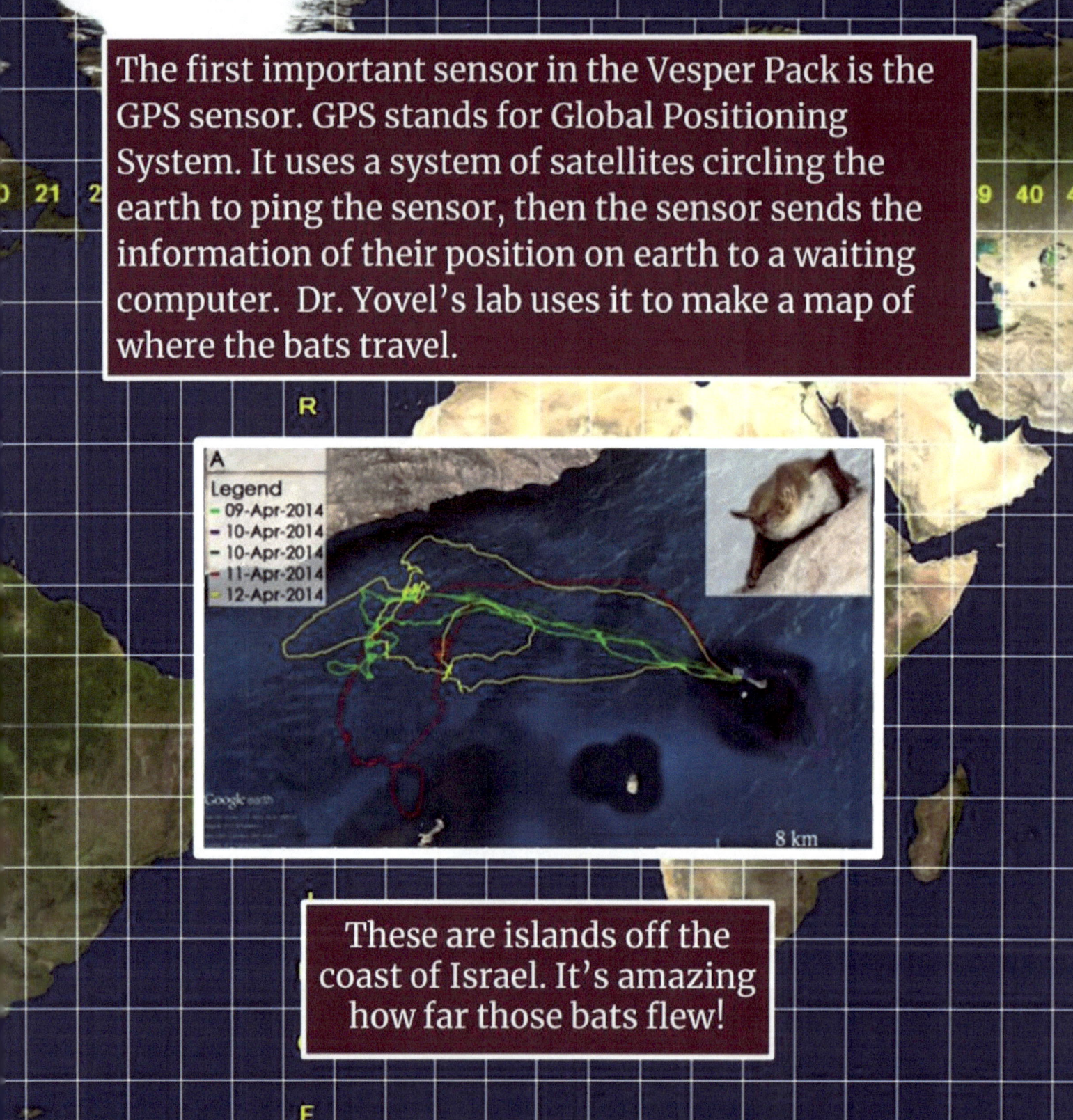

Vesper's microphones are known for being very useful with harsh environmental conditions (including dust and water). Vesper uses piezoelectric materials and MEMS technology to create the microphones.

Piezoelectricity is the electric charge that accumulates in certain solid materials in response to applied mechanical stress.

Micro-Electro-Mechanical Systems, or MEMS, can be smaller than a human hair. They are tiny machines that work with electrical current and mechanical parts.

Part of the Vesper Sensor is a microphone. This sensor can record and show us, using math, what noises the bats are making. Pretty cool, isn't it?

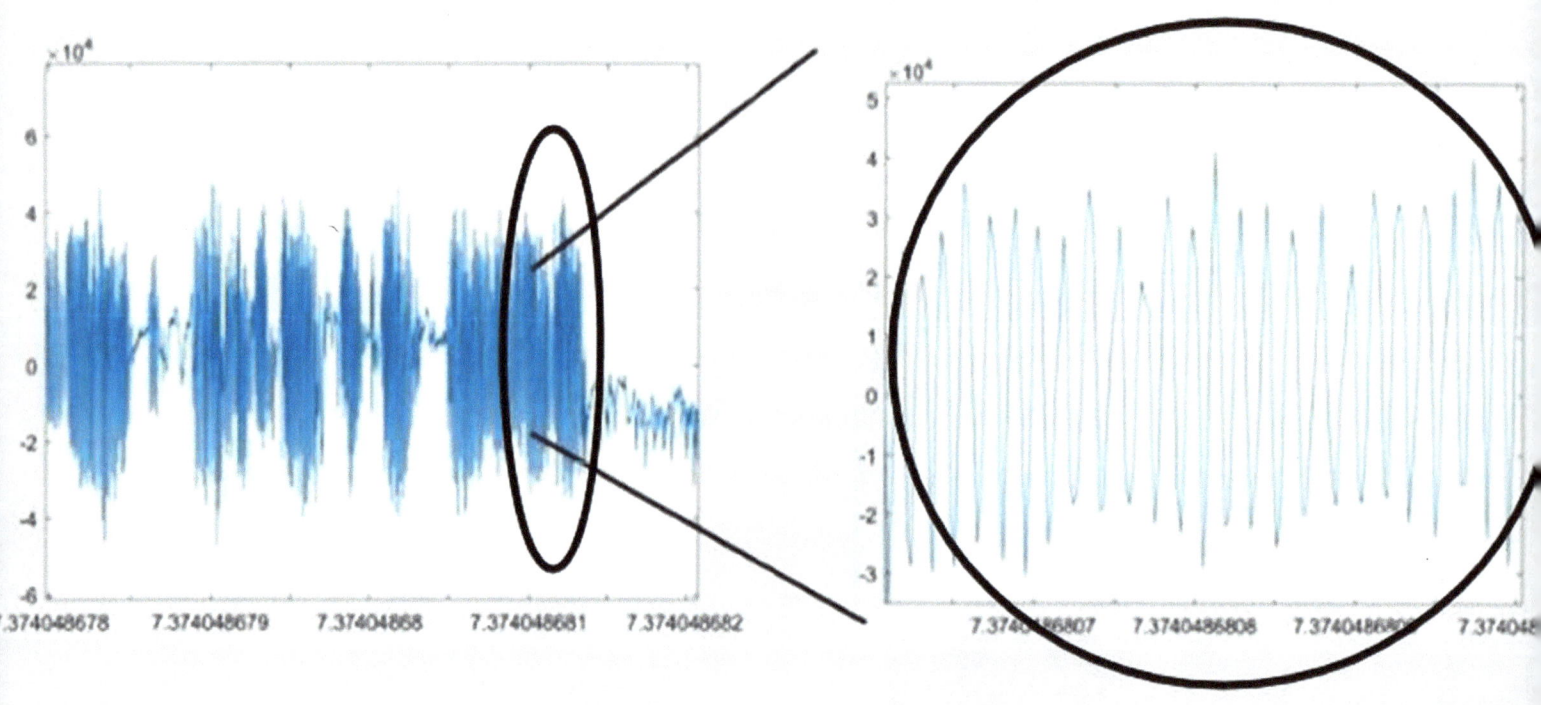

These are sound waves produced by the bats

With a computer's help, Dr. Yovel's Bat Lab is cataloging thousands of similar bat sounds.

Another sensor is called the 9DOF Sensor. It includes a compass, gyroscope, and an accelerometer. 9DOF stands for 9 degrees of freedom, that is 9 different directions of movement. A compass is north, south, east and west. A gyroscope lets you know how you are tilted. An accelerometer measures how fast you are accelerating or stopping. The 9DOF sensor combines all of these and can tell you nine different directions that the bat might be flying.

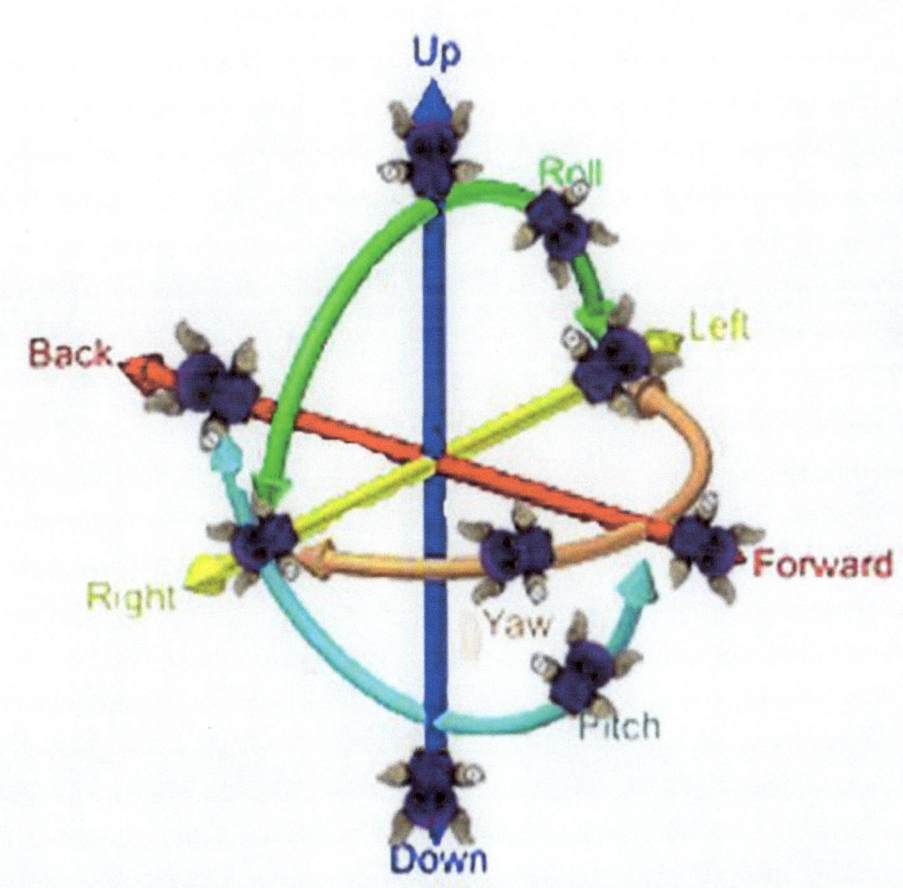

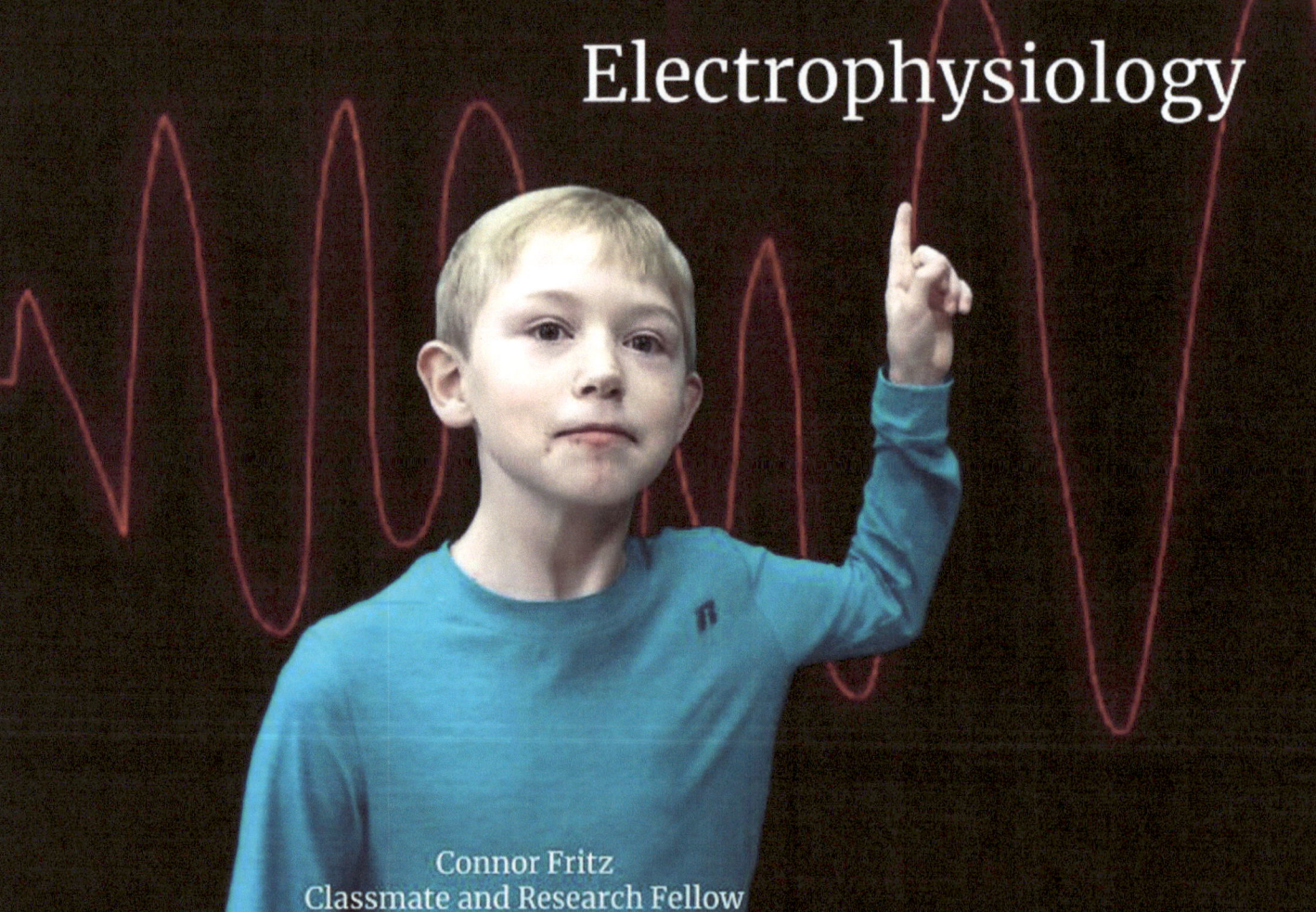

They tried to study the bats coming and going from the cave they built at Tel Aviv University. They thought they could paint numbers on the tops of their heads. However, they realized it was hard to see the numbers on camera.

They realized they could use some help. They contacted an RFID company called ReadBee and recruited the help of founder Ben Damari and their CTO Yoni Harris. The ReadBee team tested possible solutions. They created a tag that was small but could also be read when a bat flew by quickly. With our Dragon Tags, school leaders can see our history. Using similar technology, the researchers at Dr. Yovel's lab could see which bats entered and exited the University cave.

ReadBee, like the Yovel Bat Lab, is also in Israel, a country that is on the Eastern shore of the Mediterranean Sea. Israel has an extremely vibrant high-tech sector and is sometimes nicknamed the "start-up" nation.

Dragon Tags to the rescue! They used RFID technology to track which bats would return to the cave. These passive RFID tags could be attached to a collar. They integrated the RFID technology into the collar, and then attached the collar to the bat.

ReadBee installed RFID readers in a tunnel measuring 3 feet by 3 feet. Readbee installed as many as eight antennas, along with one or two readers, before settling on the most effective solution. It took a lot of testing to get it right. ReadBee also created the computer program that tracked all of the data that was used to show when the bats entered and left the cave. They were able to track which bats consistently returned to the cave.

Readbee experimented with a variety of possible tags before they developed a plastic tag, less than a centimeter wide. It was made with polyimide, a plastic that holds up under sunlight and rainwater and can handle high temperatures. The tag is connected to a copper ball band that is attached around the bat's neck.

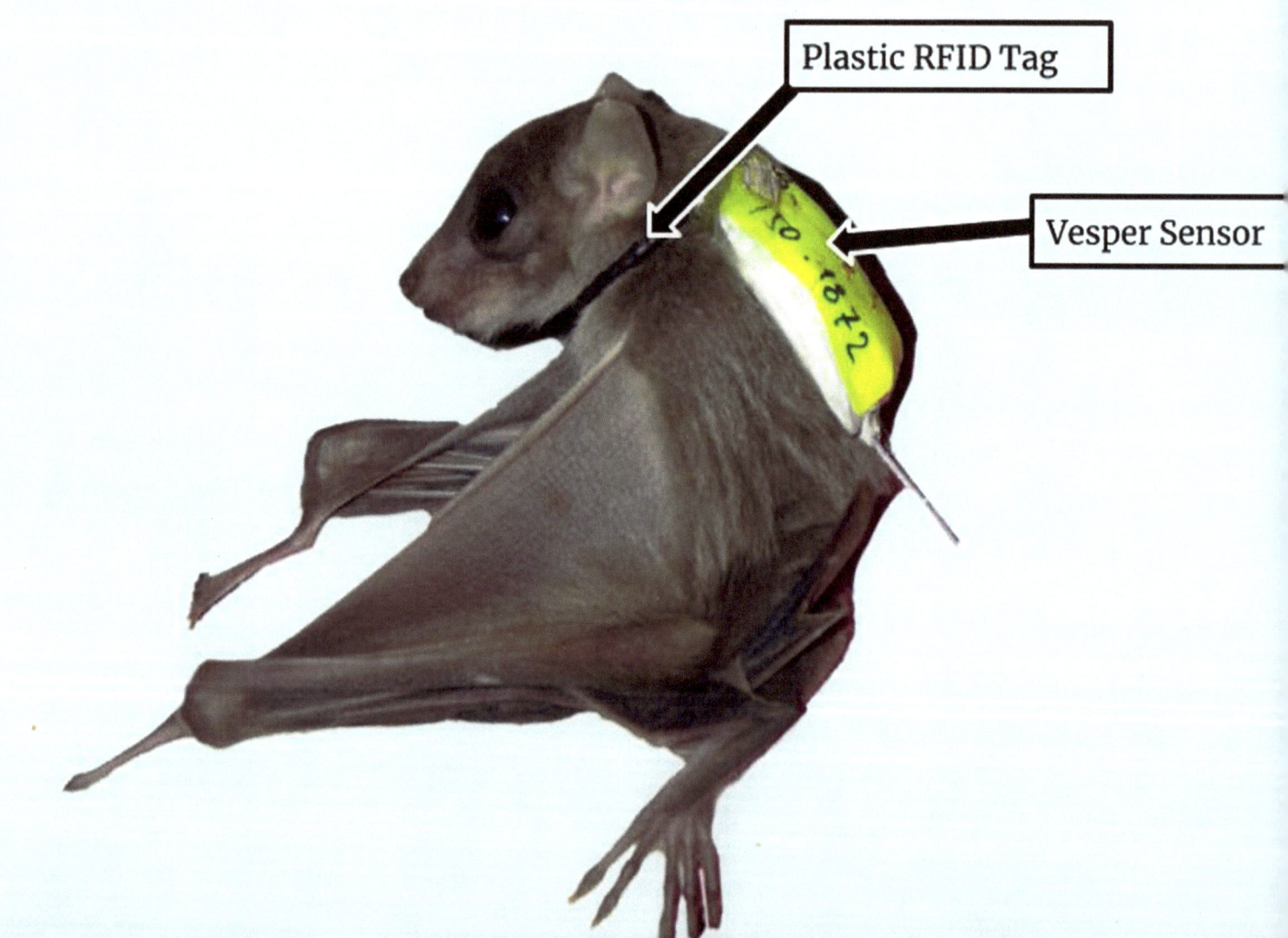

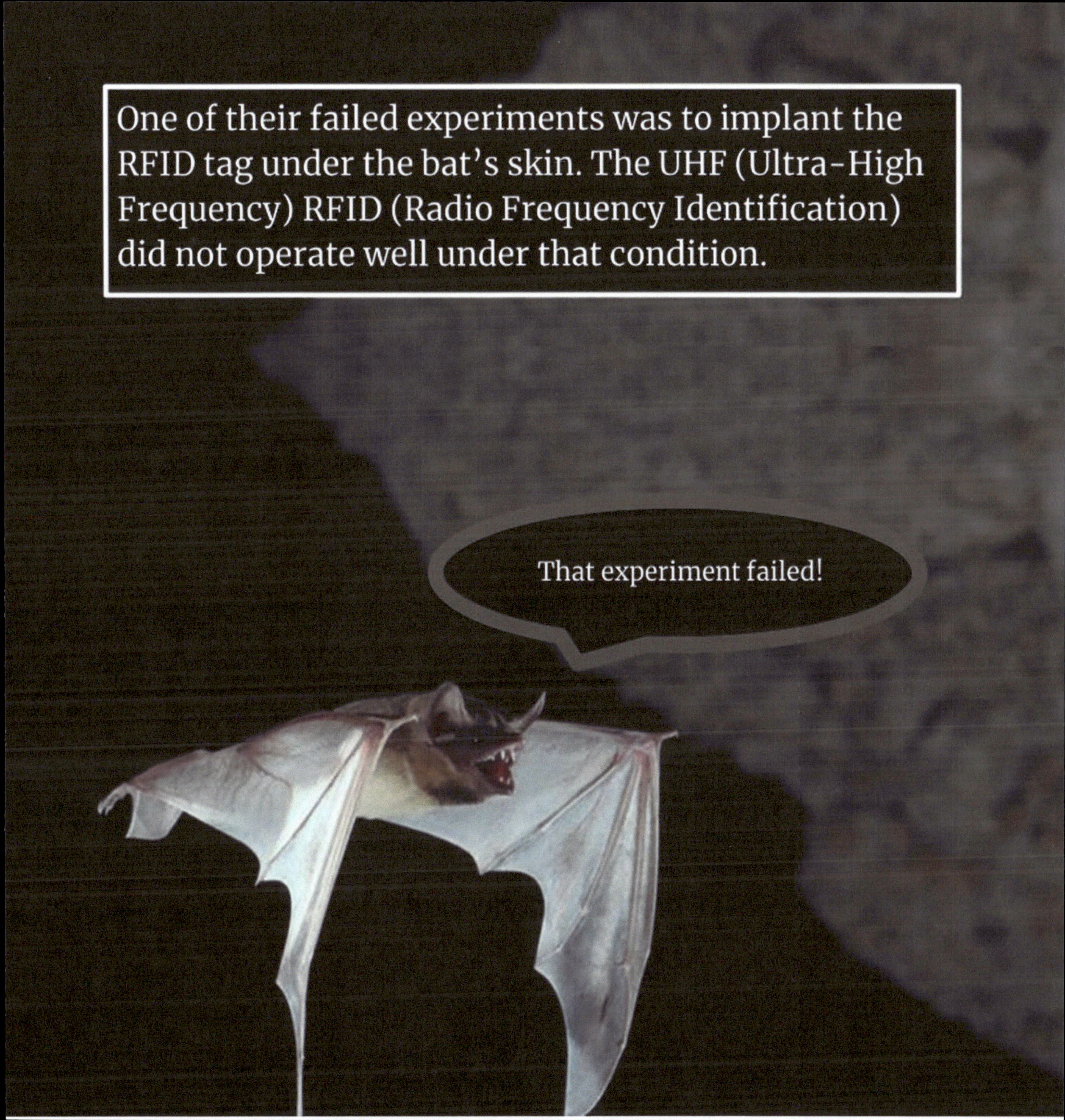

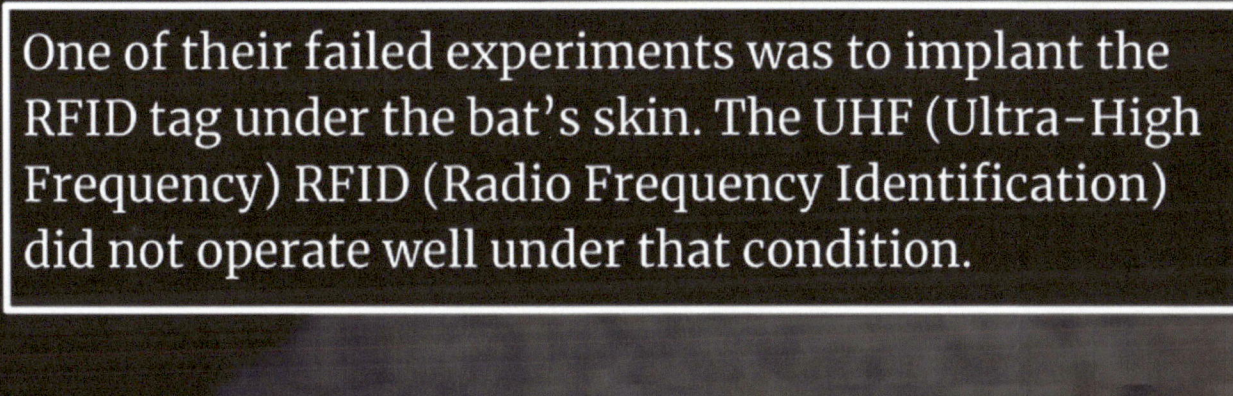

One of their failed experiments was to implant the RFID tag under the bat's skin. The UHF (Ultra-High Frequency) RFID (Radio Frequency Identification) did not operate well under that condition.

That experiment failed!

With the Vesper Sensor and the RFID tag, the Yovel Bat Lab is on the path of several new amazing discoveries.

- Discovering how the bats perform spatial navigation via echolocation.
- Learning more and more about their sensory perception.
- Observing and recording how bats behave socially.
- Tracking how the bats make decisions.

Each insight into how a bat brain works opens windows to how our brains work.

Bobby Littrell, CTO and Co-Founder of Vesper, is the recognized thought leader in piezoelectric MEMS acoustic transducers. He is both a mechanical and an electrical engineer.

Matt Crowley, CEO of Vesper, has a passion for building great teams as an industry leader. He received degrees in both physics and the philosophy of science.

Karl Grosh, CSO and Co-Founder of Vesper, is a professor of both bioengineering and mechanical engineering at the University of Michigan. He is known around the world for his research into biomechanics, cochlear mechanics, piezoelectric MEMS transducers and structural acoustics.

Bobby Littrell
CEO and Co-Founer

Matt Crowley
CEO

Karl Grosh
CSO and CO- Founder

Ben Damari and Yoni Harris are the founders of READBEE LTD. READBEE provides many RFID solutions, from tracking inventory to monitoring animal movements, all using state-of-the-art UHF hardware.

Ben Damari

Yoni Harris

The Bat Lab for Neuroecology is run by Dr. Yossi Yovel and his many University colleagues and students. They work in both neuroscience (that is the study of the brain and how it works) and ecology (how living things interact with their environment).

Dr. Yovel holds degrees in both biology and physics. He specializes in using computers to analyze behavior, how sound and signals are processed, and miniature sensors.

Dr. Yovel teaches at Tel Aviv University, including a seminar on neuroscience and courses in computational neuroscience. The work of his lab has been featured in Nova, the *New York Times*, and *National Geographic*.

batcon.org
BAT CONSERVATION INTERNATIONAL

All profits from the sale of this book go to Bat Conservation International. Their mission is the enduring protection of the world's 1300+ species of bats and their habitats and creating a world in which bats and humans successfully coexist. They are working worldwide at scale with local, regional, national and multinational public and private partners to respond rapidly and effectively to bat conservation crises, preventing the extinction of threatened bats and the extirpation of globally significant populations of bats.

Baden Academy Charter School

This public charter school in Western PA works to inspire personal excellence. They cultivate the inherent gifts and talents present in all children by providing a curriculum that integrates the arts and sciences in a highly interactive, hands-on environment.

Grow a Generation

Grow a Generation partners with gifted and talented young people and teachers to make meaningful projects possible. Faculty, students, and student teams apply in their school to be accepted into the fellowship program. Once selected, they embark on a year-long odyssey to publish a book, create a digital artifact, or enter a STEM competition. Find out more at growageneration.com